FOOD

Sophie Fern

CONTENTS

What *Is* Food? 3

Plants Make Food 4

Plant Eaters 8

Meat Eaters 9

Omnivores 11

Food Chains and Webs 12

Strange Facts About Food 14

Processed Foods 18

What *Isn't* Food? 22

Glossary 24

WHAT *IS* FOOD?

You probably think you already know what food is, but prepare yourself because this book comes with some surprises.

A simple way of describing food could be "everything you eat or drink."

Let's explore this idea. No living thing can survive without food and water. Food gives you the energy you need to grow and to repair your body. It gives you the **nutrients** your body needs to survive.

PLANTS MAKE FOOD

Animals get their energy from eating food, but did you know that most plants get their energy from sunlight?

Most plants have a bright green **substance** inside them called chlorophyll. They use chlorophyll to make their own food using energy from sunlight, along with water and carbon dioxide. The food they make is called glucose, which is a kind of sugar. We call this process photosynthesis.

Photosynthesis—how plants use chlorophyll to make their own food

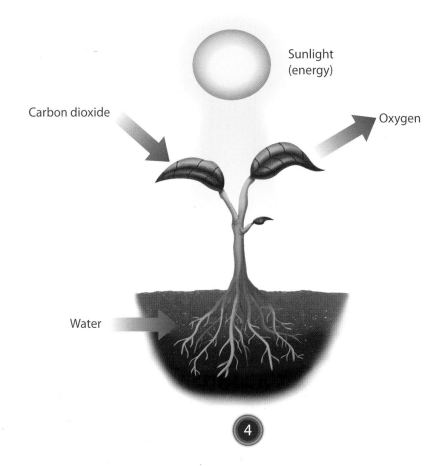

The eucalyptus is one of the fastest growing trees in the world. It can grow an inch a day! It takes a lot of energy to do this, and the energy comes from food—food the tree makes itself.

A few plants don't use photosynthesis to make their own food. They live off other plants. **Carnivorous** plants get some of their food by using their **digestive** juices outside their bodies! They then soak up the nutrients they need through their cell walls.

This plant eats other plants.

This plant eats bugs.

Fungi feed a bit like carnivorous plants. Did you know that some of the largest **organisms** on the earth are fungi? Because these fungi mainly live underground, most people don't realize just how big they are. They can have miles and miles of underground parts—called mycelia—that stretch out like branches. All you see on the surface are a few mushrooms! If you like to eat mushrooms, then you like to eat fungi.

Did you know that mushrooms are considered to be the "fruit" of fungi?

Bread is made using a fungus called yeast. As the yeast eats the sugar in bread dough it gives off a waste product called carbon dioxide. This gas is what makes the bread rise and gives it a fluffy texture. Carbon dioxide also makes the holes you see in bread!

A single spoonful of yeast contains millions of **individual** fungi.

A yeast fungus, magnified 14,000 times

PLANT EATERS

Animals cannot use photosynthesis to make their own food. They need to get their food by eating plants or other animals.

Animals that eat only plants are called herbivores. Not all herbivores live on the land. Some fish are herbivores.

An elephant spends three-quarters of its day eating. A wild adult elephant needs to eat between 220 and 440 pounds (100 to 200 kilograms) of **vegetation** a day! Dogs eat only 1 pound (0.45 kilograms) of food a day.

MEAT EATERS

Some animals don't eat plants. They eat plant eaters instead!

A food chain

Carnivores—eat other animals

Plants

Herbivores—eat only plants

Large numbers of herbivores support a smaller number of carnivores.

This jackal is waiting to eat what the lion will leave once it is full. Jackals will eat just about anything!

There are two main types of carnivores: predators and scavengers. Predators hunt and kill their **prey** before they eat it. Scavengers will eat any meat they can find, even if it's rotten.

Cat

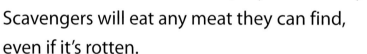

Field mouse

Grass seeds

> Did you know that cats are carnivores that *can* live on mice and water alone? By eating whole mice, cats get everything they need. They get nutrients from the vegetation in the mouse's stomach and from the meat, fur, and bones.

OMNIVORES

Some animals eat both plants and meat. They are called omnivores. Humans are omnivores because their bodies can digest both plants and meat.

> Vegetarians are people who choose not to eat animals.

FOOD CHAINS AND WEBS

A food chain is formed when plants are eaten by herbivores and then herbivores are eaten by carnivores and omnivores. When food chains are connected, they form complicated food webs.

A Prairie Food Web

Black-tailed prairie dogs

Many other plants and animals belong to this food web, too. The arrows show where the energy goes.

> Almost all the food you eat depends directly or indirectly on plants. This is true for all animals. Just think of something like a hamburger. The meat comes from an animal that eats plants. The pickle and the ketchup—they're made from plants. Everything in a hamburger except the salt is made from a plant or something that eats plants.

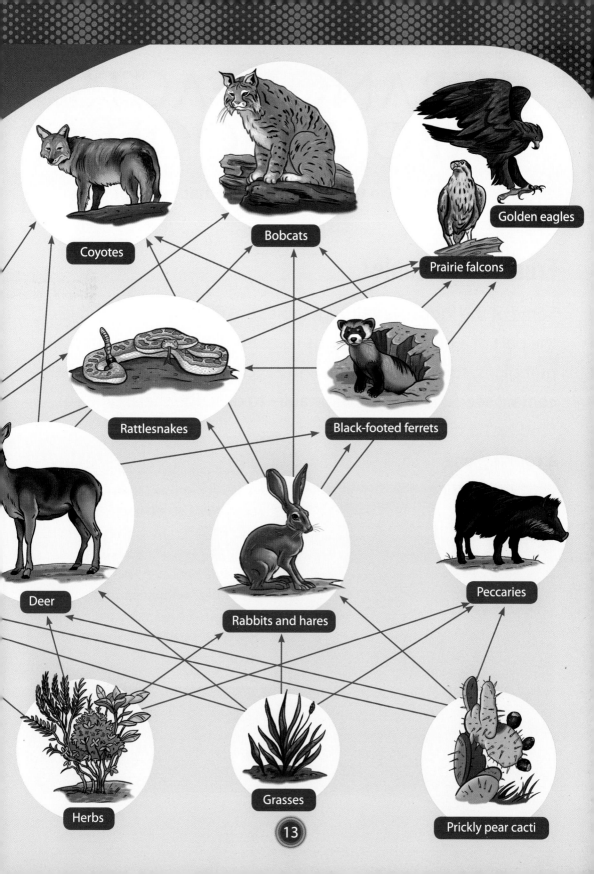

STRANGE FACTS ABOUT FOOD

There are some strange things about food that many people don't know.

Fruit or Vegetable?

Some of the foods we call vegetables are really fruits. A fruit is the part of a plant that protects its seeds. Did you know that pumpkins, squash, tomatoes, and zucchini are all fruits? They all contain seeds. The word "vegetable" to describe these types of food isn't a scientific term, but a cooking one.

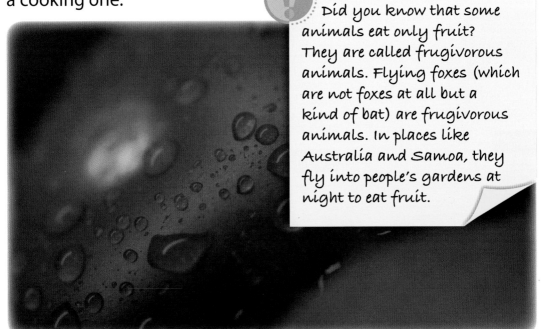

Did you know that some animals eat only fruit? They are called frugivorous animals. Flying foxes (which are not foxes at all but a kind of bat) are frugivorous animals. In places like Australia and Samoa, they fly into people's gardens at night to eat fruit.

Eating Grass

People don't usually eat the leaves of grass because they are hard for us to digest. We eat things made from grass stems and seeds, like sugar, corn, and rice. Chopsticks are often made out of bamboo, and bamboo is a grass. When you eat rice with chopsticks, you are eating a grass with a grass!

Plant eaters like cows have four-chambered stomachs that help them digest grass. As the grass travels through the different chambers, bacteria break down the grass and makes it digestible.

> Did you know that milk chocolate bars are made of seeds and grass? It's true! Chocolate is made from cacao seeds and sugar cane. Cacao seeds grow on plants. Sugar cane is a kind of grass.

> Rabbits have only one stomach, just like people, but they eat lots of grass. The grass, along with its nutrients, mostly passes right through rabbits. To get all the nutrients they can get from grass, rabbits eat their first droppings. The hard little pellets you see on the ground come only after the rabbits have eaten the grass twice!

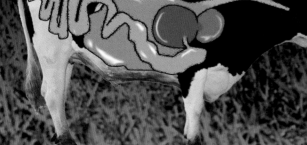

A cow's four-chambered stomach

Waste Products

Just as every living thing needs food, every living thing also gives off waste products. Living things give off different waste products based on what they need and do not need to survive. The waste product of photosynthesis is oxygen, which we breathe.

> People are not the only living things that use the waste product of something else. Dung beetles lay their eggs in balls of animal **dung**. When the baby beetles hatch, they eat the dung.

Bugs

Bugs are food for some animals—and for some people, too! Would you like a fried spider for lunch? Fried spiders are popular

Fried spiders

in Cambodia and some parts of South America. They are high in protein— the perfect snack! In some countries, people like to eat deep-fried grasshoppers.

> People also eat bugs by accident. It has been estimated that the average person eats about 2.2 pounds (1 kilogram) of insect pieces in their food every year!

Some animals, such as shrews, mostly eat bugs. These animals are called insectivores. Even if you don't like eating bugs, you

have probably eaten something bugs make. Honey is **regurgitated** bee spit. Bees go from flower to flower and collect **nectar**. When they get back to the beehive, they regurgitate the nectar into a honeycomb as honey.

PROCESSED FOODS

Fresh food is tasty and good for you, but people can't always eat it all before it starts to **decay**. Some of our food is seasonal, which means that we can harvest it only once a year. Since we like eating these things all year round, scientists have found ways to keep food **edible** for longer.

The bacteria that spoil food like living in warm, wet conditions, with oxygen and light. By removing any one of these, we can slow down the bacteria. Canning is one way to keep food from spoiling. Freezing, drying, and pickling are some other ways.

We actually like some foods that are not really fresh. For example, you probably like moldy milk on your pizza. But you call it cheese!

Drying

Drying food is a good way to make sure that it stays safe to eat. Drying removes the moisture. Bacteria can't survive in really dry food because they don't have enough water. All living things need water to survive—even bacteria!

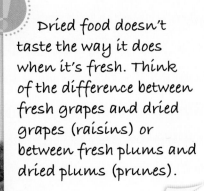

Dried food doesn't taste the way it does when it's fresh. Think of the difference between fresh grapes and dried grapes (raisins) or between fresh plums and dried plums (prunes).

Freezing

Bacteria can't **multiply** as fast when it's cold, so cold food stays safe to eat longer than food that's been kept at room temperature. This is why we put food in the refrigerator. The very cold conditions of a freezer stop the bacteria in food from multiplying at all.

Russian scientists have found frozen woolly mammoths in the Arctic that died thousands of years ago. They tried cooking a little bit of the meat and found that it was still safe to eat.

Food for Astronauts

When you take things into space, weight really matters. It takes a lot of fuel just to take one pound (0.45 kilograms) into space, so scientists have found a way to make food very, very light. It's called freeze-drying.

Here is how it's done. A frozen block of ice cream, for example, is put into a **vacuum chamber**. This removes the air and water and leaves the ice cream light and dry. Just add water, and you can eat it.

The most important food **utensil** in space is a pair of scissors. Astronauts need scissors to open the packages their freeze-dried meals come in!

WHAT *ISN'T* FOOD?

We don't call things food if they are not safe to eat. Some kinds of mushrooms are safe to eat. We think of *them* as food. Other kinds are poisonous, so we don't eat them. We don't think of *them* as food for humans.

People can't safely eat this mushroom, but some insects can.

There are things we can eat without harm but don't think of as food.

We also eat some things by accident, but we don't count them as food.

Everyone eats a little bit of dirt each year by accident. It usually doesn't cause any harm.

There are also things that are safe to eat only some of the time. We think of them as food only when they are safe to eat. In California, for example, people are careful about eating shellfish. They watch for red tides, which are a sign that the shellfish are eating poisonous algae. If this happens, people stop eating shellfish until they are safe to eat again.

Some people can't eat the foods that other people can eat safely. For example, some people are allergic to foods other people can eat. If they do eat them, they get sick.

We also talk about foods that are not good for us. What this means is that, although our bodies can digest these foods, eating too much of them isn't healthy.

Food, then, is different things to different people, animals, and plants. What food is depends on who eats it!

GLOSSARY

carnivorous—animal-eating

decay—decompose

digestive—able to break down food

dung—manure

edible—able to be eaten

individual—single

multiply—increase

nectar—the sweet juice in a flower that a bee turns into honey

nutrients—the things that provide energy and raw materials to help an organism grow, repair itself, and stay alive

organisms—living things, such as plants and animals

prey—animals that predators eat

regurgitated—thrown up

substance—the material that something is made of

utensil—a tool used for eating or cooking

vacuum chamber—a machine that sucks all the air out

vegetation—plant material